ENDANGERED ANIMALS AROUND THE WORLD

ENDANGERED MAMMALS AROUND THE WORLD

BY GOLRIZ GOLKAR

PEBBLE
a capstone imprint

Published by Pebble, an imprint of Capstone
1710 Roe Crest Drive, North Mankato, Minnesota 56003
capstonepub.com

Library of Congress Cataloging-in-Publication Data is available on the Library of Congress website.

ISBN: 9780756578312 (hardcover)
ISBN: 9780756578404 (paperback)
ISBN: 9780756578411 (ebook PDF)

Summary: From towering elephants to lynxes with tufted ears, these endangered mammals are having a tough time. Learn about some incredible mammals that need our help to survive.

Editorial Credits
Editor: Ericka Smith; Designer: Sarah Bennett; Media Researcher: Svetlana Zhurkin; Production Specialist: Katy LaVigne

Image Credits
Getty Images: astepforever, 11, Dennis Stogsdill, 13, Jose Angel Cortes Garcia, cover, Kerry Hargrove, 8, Nick Everett, 20; Newscom: Polaris/Rod Harbinson, 18, Sipa USA/Alex Milan Tracy, 9; Shutterstock: aDam Wildlife, 16, Animanish, 26, asantosg, 7, BAZA Production, 29, JB Frames, 21, JHVEPhoto, 27, kusari, 24, nutsiam, 22, Pato Lucas, 15, Paulphin Photography, 4, PhotocechCZ, 25, Prasun Kumar Chakraborty, 23, structuresxx, 5, tacud, 10, tony mills, 14, Travel Stock, 6, Viacheslav Lopatin, 1; Superstock: age fotostock/Sergi Reboredo, 17, Hemis/Cegalerba-Szwemberg, 19

Printed in the United States 6601

TABLE OF CONTENTS

Words in **bold** are in the glossary.

All About Endangered Mammals

What Is a Mammal?

Dolphins splash along a coast. They squeak at one another and leap in the air. Dolphins are mammals.

Mammals are **warm-blooded** animals with hair or fur. They produce milk for their young. Most mammals give birth to live babies. A few **species** lay eggs.

There are more than 5,000 known mammal species. They include humans, bears, monkeys, cows, whales, kangaroos, bats, and rodents. They are found on every continent.

What Is an Endangered Mammal?

About one in five species of mammals is **endangered** or **extinct**. Some species are close to dying out. Others are no longer alive.

Why are some mammals close to dying out? Farming and construction destroy their **habitats**. **Climate change** makes some places too warm for mammals to survive. Pollution and disease are big problems too.

Where Do Endangered Mammals Live?

Endangered mammals live all over the world. Many are **native** to Asia.

Here's where you can find the mammals you'll learn about in this book!

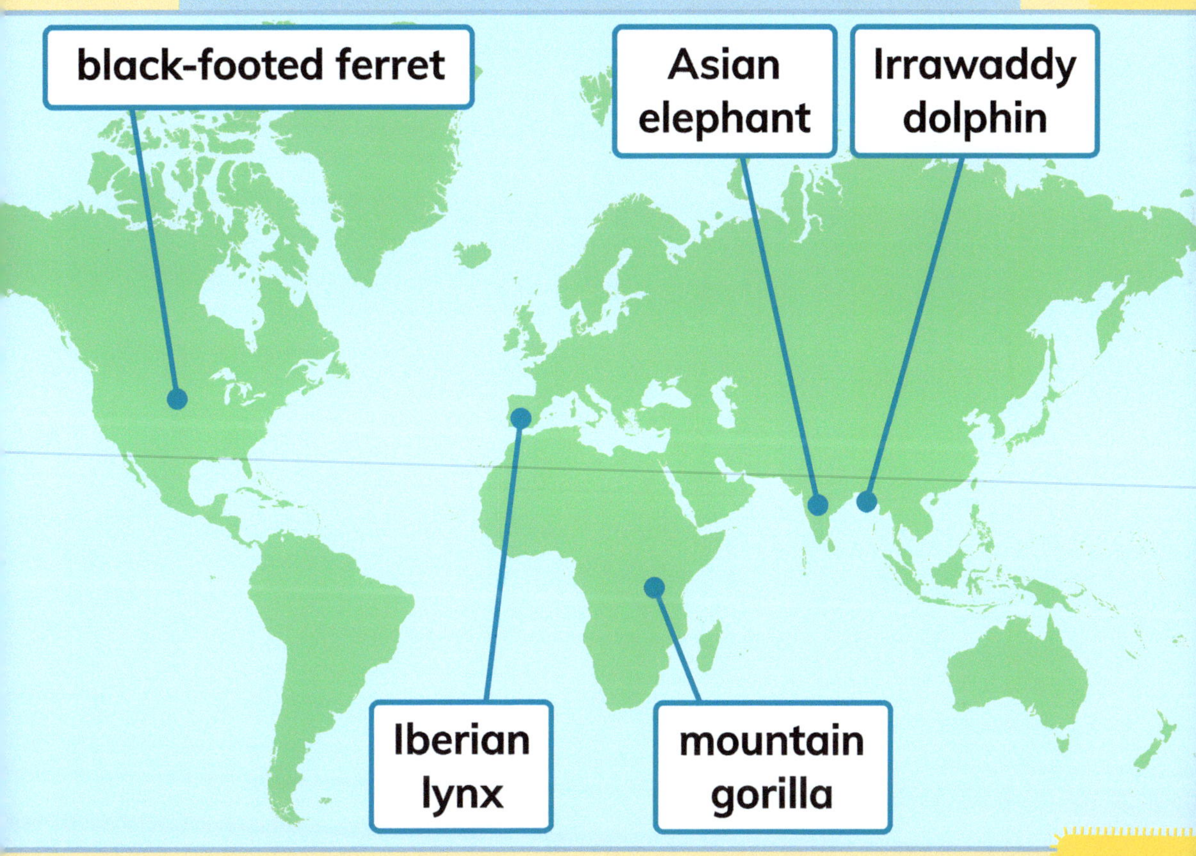

Black-Footed Ferret

A black-footed ferret scurries across a **prairie**. It spots a prairie dog. *Yum! Time to eat!*

Black-footed ferrets are endangered for many reasons. When people clear prairies for farming, ferrets lose their habitat. Farmers also kill prairie dogs. Then, the ferrets have less food. Ferrets live in **burrows** dug by prairie dogs. So they also lose their homes. Disease is a probem too.

Groups of people are helping black-footed ferrets. They **breed** them. Then they release them into the wild.

Mountain Gorilla

Just over half of all mountain gorillas live in a protected area in the Virunga Mountains in central Africa. The other half live in the Bwindi Impenetrable National Park in Uganda.

Cattle grazing near Bwindi Impenetrable National Park

Mountain gorillas are endangered. Farming, charcoal **harvesting**, and raising livestock destroy their habitat. By the 1980s, there were fewer than 400 of them in the wild.

Other human activity is a problem too. People fight wars and scare gorillas away. The gorillas must move farther north to much colder areas. Mountain gorillas also get sick from humans. And they get stuck in traps set for other animals.

But things are changing. Hunting is banned in protected areas to prevent trapping. People are cutting down fewer trees to save the gorillas' habitats. And locals are trying to end wars on gorilla land. Now, more than 1,000 mountain gorillas live in the wild.

Iberian Lynx

The Iberian lynx is native to the Iberian Peninsula. Its population started to drop in the 20th century. It mainly eats rabbits, but rabbits are dying from diseases. Lynxes are hit by cars traveling on roads built in their habitat. Many are hunted illegally. And some get caught in traps by accident.

A lynx that will be released into the wild

Groups are working to protect the lynx's habitat. They also breed the lynx and release the adults into the wild.

Irrawaddy Dolphin

A dolphin with a round face pokes its head above the water. It spits water into the air! It's playing with its friends.

An Irrawaddy dolphin in the Mekong River in Cambodia

Irrawaddy dolphins live in coastal areas of southern Asia. They also swim in Asian rivers. These dolphins are important to their habitat. They eat fish, which helps keep the river fish population balanced.

But Irrawaddy dolphins face many problems. They are caught in fishing nets by accident. They can get hurt or die. People also catch the dolphins to sell them. People build dams that block their access to rivers. And farming and construction pollute their waters.

An Irrawaddy dolphin in the Philippines

People are helping these dolphins. They are improving fishing practices. They are also working to stop people from catching and selling the dolphins.

Asian Elephant

An Asian elephant munches on grass. Its trunk squirts water into its mouth.

Asian elephants are the largest land mammals in Asia. They spend most of the day eating plants. They always stay near water. They need lots of water to survive.

Asian elephants are an important animal in their habitat. They spread seeds while eating. This helps new plants grow. They also create paths in forests. Other animals use the paths.

But Asian elephants are endangered. Farming and mining destroy their habitat. Cities and roads are built on their land. The elephants are forced into small forests nearby. Buildings block their migration, so they cannot travel to find food and water.

Some Asian elephants are **poached** for their meat or tusks. Some kill each other in fights over food.

Groups are now working to protect their habitat and to stop poaching.

Making Progress

Indian Rhino

The Indian rhino lives in northern India. It helps the local **ecosystem.** The rhino eats tall grasses and shortens them. Then other animals can eat them more easily. Rhinos also drink lots of water. Rain refills those water sources. This keeps the water clean.

In the 1970s, the population of Indian rhinos dropped. They were poached and lost their habitat.

People have helped them come back. Their habitat is now protected. Poaching has decreased too.

West Indian Manatee

West Indian manatees have a wide range. They swim along the coasts of Florida and down to South American waters.

These sea animals were once endangered. They were hunted. Many were struck by boats. Nearby construction caused water pollution, and toxic plants started to grow. Manatees could die from eating them.

Groups worked to stop people from hunting the manatee. They teach people about reducing pollution. The West Indian manatee is slowly coming back.

How You Can Help

Endangered mammals need help to survive. Here's what you can do:

- Help keep their habitats safe by not polluting water and soil with trash.
- Do not use chemicals at home that can harm animals and the environment.
- Help clean up the environment.
- Use less water.
- Spread the word!

GLOSSARY

breed (BREED)—to mate and produce young

burrow (BUHR-oh)—a hole in the ground made or used by an animal

climate change (KLY-muht CHAYNJ)—a significant change in Earth's climate over a period of time

ecosystem (EE-koh-sis-tuhm)—a group of animals and plants that work together with their surroundings

endangered (en-DAYN-juhrd)—in danger of dying out

extinct (ek-STINGK)—no longer living

habitat (HAB-uh-tat)—the home of a plant or animal

harvest (HAR-vist)—to collect or gather

native (NAY-tuhv)—living or growing naturally in a place

poach (POHCH)—to hunt or fish illegally

prairie (PRAIR-ee)—a large area of flat or rolling grassland with few or no trees

species (SPEE-sheez)—a group of animals with similar features

warm-blooded (WARM-BLUHD-id)—having a body temperature that stays about the same all the time

READ MORE

Chanez, Katie. *Manatees*. Mendota Heights, MN: Apex Editions, 2022.

Jaycox, Jaclyn. *Unusual Life Cycles of Mammals*. North Mankato, MN: Capstone, 2021.

Kington, Emily. *Gorillas*. Cornwall, UK: Hungry Tomato, 2022.

INTERNET SITES

Kiddle: Mammal Facts for Kids
kids.kiddle.co/Mammal

NASA Climate Kids: A Guide to Climate Change for Kids
climatekids.nasa.gov/kids-guide-to-climate-change

National Geographic Kids: Habitat Destruction
kids.nationalgeographic.com/science/article/habitat-destruction

INDEX

ABOUT THE AUTHOR

Golriz Golkar is the author of more than 70 books for children. A former elementary school teacher, she holds degrees in American literature and literacy education. Golriz enjoys reading, cooking, singing with her daughter, and looking for ladybugs on nature walks.